KNOWLEDGE ENCYCLOPEDIA
CONSTELLATIONS
SPACE

© Wonder House Books 2023

All rights reserved. No part of this book may be reproduced or transmitted in any form by any means, electronic or mechanical, including photocopying and recording, or by any information storage and retrieval system except as may be expressly permitted in writing by the publisher.

(An imprint of Prakash Books)

contact@wonderhousebooks.com

Disclaimer: The information contained in this encyclopedia has been collated with inputs from subject experts. All information contained herein is true to the best of the Publisher's knowledge. Maps are only indicative in nature.

ISBN : 9789390391509

Table of Contents

The Amazing World of Constellations	3
Constellations in the Sky	4–5
A Brief History	6–7
Spotting Constellations	8–9
Star Maps	10–11
Constellations in the Northern Hemisphere	12–13
Constellations in the Southern Hemisphere	14–15
Finding Stars and Constellations	16–17
The 'C' Constellations	18
Legends of the Horses	19
A Tale of Two Giants	20
Water Creatures	21
The Zodiac	22–24
What is Astrology?	25
Chinese Constellations	26–27
Stargazing Apps and Astrotourism	28–29
Official Naming of Constellations	30
NASA's New Constellations	31
Word Check	32

THE AMAZING WORLD OF CONSTELLATIONS

*'Over the edge of the World now comes forth
Great Orion...
Hunter of the Stars...
Behold the gleaming star-fire of his sword!'*
—Henry Wadsworth Longfellow, American poet

The poem refers to one of the famous star groups or constellations named after a Greek mythological character—Orion the Hunter. The symbolism and imagery in the poem reflect the importance given to stars and constellations in our culture. For centuries, people from the ancient and the modern world alike have been inspired by the stars and studied the constellations. They did this across the boundaries of countries and cultures.

Constellations were first studied for religious reasons. People believed that stars and constellations revealed important messages and stories about their Gods and Goddesses. Today, constellations are helpful to astronomers in identifying the stars located in the sky. They also provide hours of pleasure to stargazing enthusiasts.

▼ *Stargazing helps us appreciate the vastness of the universe*

Constellations in the Sky

Since ancient times, human beings have gazed at the stars and looked to them for help in navigating across the mighty oceans. Stars were also used to predict the seasons and prepare plans for sowing seeds and harvesting crops. In order to easily recognise and gather information from this 'celestial calendar' in the skies, stars (brighter ones in particular) were grouped together as constellations and named according to their apparent shapes.

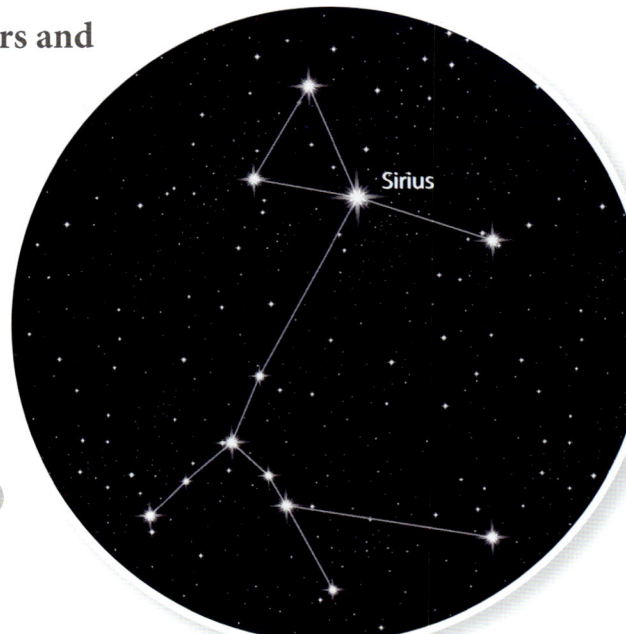

▲ *The Canis Major constellation with Sirius, its brightest star*

 ## What is a Constellation?

In astronomy, a constellation is defined as a group of stars. They are not exactly real; in order to easily pick out the different stars in the night sky, people made up patterns of stars and named these patterns according to the apparent shapes that they took. These patterns might display objects, creatures or mythical people. So, these star groups or constellations are named after those apparent shapes. For example, there are groups of stars which seem to take the shape of a big bear or a lion and a hunter.

 ## Importance of Constellations

Constellations helped in locating stars and mapping the night sky. For example, if you know enough about constellations, you might spot Orion one evening, by identifying three of the bright stars that form the Hunter's belt. Remember that Canis Minor and Canis Major are nearby. Without constellations, finding a speck of a star in the vastness of the night sky would be very difficult.

Constellations are also known to play a role in farming, agriculture and navigation. Earlier, when there were no calendars, constellations and stars were the only way to identify the harvesting and sowing seasons. Besides, the position of the North Star allowed travellers to locate their exact location and travel across the globe.

 ## How many Constellations are There?

Officially, there are a total of 88 constellations, out of which 48 were identified by the ancient Greeks. Since some of the main constellations include more than one form or creature, the total number of shapes and images formed in these constellations is more than 88 and includes 9 birds, 2 insects, 19 mammals, 10 water creatures, 1 constellation shaped like a historical figure's hair, a serpent, a flying horse, a dragon, 2 centaurs (a mythological creature that is half-man and half-horse), 1 river, 29 inanimate objects and 14 men and women.

 ## Christening the Constellations

In ancient times, people named constellations after mythological creatures or characters, or even objects. Some were named after religious beliefs. Most of these star patterns do not exactly represent the names that were given to them. The names are suggestive. They were meant to be indicative, not representative.

▲ *Ursa Major or the Great Bear constellation. Ancient Greeks associated the constellation with the nymph Callisto, who was positioned in heaven by Zeus in the form of a bear*

| SPACE | CONSTELLATIONS |

APPROXIMATELY 20°

▲ *The Big Dipper as measured by its angle or distance between the two stars at the end of the constellation*

Measuring Constellations

When we discuss the size of a constellation, we are referring to how big it looks to us from Earth. But how much space does it occupy in the sky? Generally, astronomers use angles to measure the size and distance of a constellation. As an experiment, if you have two sticks which are joined together at one end, spread them apart so that each stick is pointing to the stars at the two ends of a constellation. Measure the angle in between them to get the angular distance between the two stars at the ends of the constellation. This is not an exact or accurate measurement. Take the example of the constellation shown in the diagram. You can see that the angle between the two lines is approximately 20°, which is the size of the Big Dipper, a bowl-shaped constellation.

Incredible Individuals

Friedrich Wilhelm Bessel (1784–1846) was a talented man with many astronomical accomplishments. Despite being born into a poor German family, he pursued his dreams and interest in travel, languages and navigation. This led him to astronomy and mathematics. Impressed by his 1804 paper on the Halley's Comet, German astrologer Wilhelm Olbers recommended that he join the Lilienthal Observatory. Bessel had to make a tough decision—he could either stay with the company that had employed him and make a good living, or pursue a career in astronomy but remain in poverty. He chose the latter!

Bessel made many important contributions, including noting the precise measurements or positions of approximately 50,000 stars with his meticulous methods of observation. He thus established a new level of precision in astronomy. He was responsible for building Germany's first big observatory in Konigsberg.

▲ *Friedrich Wilhelm Bessel*

A Brief History

Who were the first people to think of or imagine the star patterns in the sky? Nobody can tell when and where constellations originated. Once, there was a widespread belief that the ancient Greeks were responsible for the naming and identification of constellations. However, this belief has now been disproved. So, who exactly is responsible for coming up with constellations?

⭐ Who Made them Up?

The earliest systematic detailing of constellations was found in the poem, *Phaenomena*, written by a Greek poet named Aratus in 270 BCE. He described 43 constellations and named five stars. However, the system of naming and labelling constellations had started long before this. It was devised by the ancient Babylonians and Sumerians. The information about the constellations travelled to Egypt and was picked up by ancient Greeks, who began writing about them. The ancient Greeks were responsible for our contemporary system of constellations.

⭐ Major Contribution of the Greeks

While it is untrue that the ancient Greeks came up with the system of constellations, they were responsible for finding and naming many of them. Ptolemy was an Egyptian astronomer and mathematician of Greek origin who published the *Almagest* in 150 CE. Besides other astronomical information, it included a catalogue of 1,022 stars organised into 48 constellations. This was the most important list of constellations and stars. It laid the foundation of the modern constellation system. The remaining 40 constellations, of the total 88, were contributed by a few European astronomers.

▲ *This picture is from the Epitome of the Almagest (1496), and shows Ptolemy and Regiomontanus. During the Renaissance, the book was one of the most important sources on ancient astronomy*

◄ *Claudius Ptolemy, the famous Egyptian astronomer and mathematician*

How were Stars Named in Constellations?

The Greeks named and referred to stars by their location in a constellation. However, this was not always practical and could be quite a complicated practice. While describing the Aldebaran star, for example, Ptolemy mentions it as 'the reddish one on the southern eye'. In the 10th century, new star names were introduced by Al-Sufi, a well-known Arabic astronomer, in his version of the *Almagest*.

Nomadic Arabs often gave names to bright stars, for example Aldebaran and Betelgeuse (some were translated from Ptolemy's book). Much later, the works of Ptolemy and other Greek books were translated from Arabic into Latin and again introduced in Europe by Arabs. We are more familiar with Ptolemy's work due to the Arabic translation of his book into Latin. Therefore, we have a Greek system of constellations with Latin names and stars that have Arabic names!

▲ *This illustration of a decorative border depicts the Greek myths associated with the last six signs of the zodiac*

 ## Other Cultures and Constellations

People from many cultures and countries have visualised patterns in the stars, however, they often interpreted or saw them differently. For example, the Big Dipper is an asterism. An asterism is a smaller pattern of stars that might belong to a bigger constellation. Sometimes an asterism has stars that belong to more than one constellation. The Big Dipper is an asterism or a small pattern of seven stars, which is a part of the larger Ursa Major constellation. Different cultures tell different stories about them. The British seem to visualise it as the Plough. In the South of France, they call it a saucepan (la casserole). In Ancient India, the stars were referred to as the 'Saptarishi', or seven wise men and so on.

Incredible Individuals

French astronomer Nicolas Louis de Lacaille (1713–1762) is best known for mapping the constellations which are seen from the southern hemisphere. He is responsible for naming several of them. He was a professor of mathematics in Paris, but in 1741, he joined the Academy of Sciences. During 1750–1754, he headed an expedition to the Cape of Good Hope in South Africa. Here, within a short span of two years, he was able to pinpoint the positions of about 10,000 stars—many of them are referred to using the numbering system in his catalogue, even today. He returned to France in 1754 and worked alone on all the data he had gathered. It seems overwork and exertion were responsible for his death. In 1763, his *Coelum Australe Stelliferum* or *Star Catalogue of the Southern Sky* was published.

▲ *Nicolas Louis de Lacaille*

Spotting Constellations

Some constellations are seasonal, but some can be seen year-round and appear to be fixed in place. The constellations that people can view also depend on the place and time when they are being viewed. Do you ever wonder why this is the case and why constellations appear to rise and set?

Which Constellations can you Spot?

▲ An old illustration depicting the constellations visible from the northern hemisphere and the southern hemisphere

People in different parts of the world are able to view different constellations, depending on whether they live above the Equator or below it. All places above the Equator are part of the northern hemisphere, and all places below it are part of the southern hemisphere.

People living above the Equator will only be able to see the constellations that appear in the sky above the northern hemisphere. Similarly, people living below the Equator will only be able to see the constellations that appear in the sky above the southern hemisphere. If you live near the Equator, then you will be able to view some constellations from both the hemispheres.

Why do Stars Seem to Move Across the Sky?

Sometimes, stars seem to move across the night sky. To understand why it appears so and why everyone on Earth cannot see all constellations, we need to understand Earth's motions. These are Earth's rotation on its own axis and its revolution around the Sun.

As Earth rotates, one side of it faces the Sun and the other side faces away from the Sun. So, the side facing the Sun receives light. This causes day. The side facing away from the Sun does not receive light. This causes night. Though stars are always present in the sky, they are visible only at night, when Earth is facing away from the Sun's glare.

Earth's revolution around the Sun causes the seasons, like monsoon, summer and winter. Both aspects—the seasons and night and day, cause people in different parts of Earth to view different parts of the night sky during different seasons. This makes people think that stars are moving when, in fact, it is Earth that is moving.

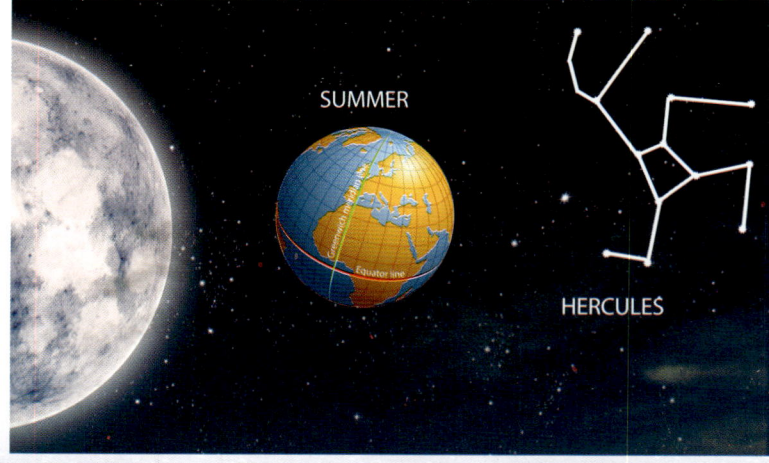

▲ Earth's rotation and revolution make stars appear to move across the skies, and hence not everyone can see all constellations all the time

Circumpolar Constellations

Due to Earth's movement around its axis and around the Sun, most stars and constellations seem to rise and set. Hence, they shift in the night sky. However, in the northern hemisphere, a few constellations close to the North Star (Polaris), like Ursa Major, Ursa Minor and Cassiopeia, are visible in the night sky on most days of the year. Even though they appear to spin a full 360° around Polaris during the night, as Earth rotates around its axis, they will still always be visible above the horizon. These are known as **circumpolar constellations**. The stars farther away from Polaris will be visible above the horizon only for some part of the night, as they rotate in wider circles. In the southern hemisphere, the equivalent circumpolar constellations are Crux (Southern Cross) and Carina.

▲ The pictures show two circumpolar constellations of the southern hemisphere, Crux and Carina

▲ Circumpolar constellations in the northern hemisphere rotate around Polaris and can be viewed all through the year

In the northern hemisphere, constellations such as the Summer Triangle appear only in the summer season. This is because the Sun's glare makes them difficult to view during winters.

▲ The Summer Triangle in the northern hemisphere: Lyra or the Harp constellation with Vega, its brightest star; Aquila or the Eagle constellation with Altair, the brightest star which forms the eye of the Eagle; and Cygnus or the Swan constellation with star Deneb in its tail

Isn't It Amazing!

For stargazing enthusiasts and astronomers, the year 2022 will be historic. For the very first time, scientists have been able to predict the birth of a new star. It will be born in the Cygnus constellation, which is the most visible one in the night sky. This extraordinary astronomical event is going to be a once in a lifetime experience, as people will be able to witness it without a telescope.

Star Maps

▲ An ancient star map

Star maps help astronomers and stargazers easily locate stars, constellations and other celestial bodies. There are different types of maps. In ancient times, globes were more commonly used for the purpose. Star maps were also drawn, carved and painted. Present-day maps are based on a system similar to geographic latitudes and longitudes. These maps are created from photographs taken by Earth-based equipment, or from images from satellites and spacecrafts.

 ## What is an Astronomical Map?

An **astronomical map** is a scientifically drawn map of stars and galaxies, but it also includes planets and their moons in the night sky. A typical map shows the relative positions of the stars and their brightness. To accurately locate stars and constellations, you need to have the correct map, depending on where you are located and the season when you are observing them.

 ## How to Find the Appropriate Map?

Generally, there are four different star maps—one for each season—winter, summer, spring and autumn. The time of observation at night is also important. This is because stars are not stationary. In addition, Earth's revolution and rotation are factors that make it seem like the stars are moving across the sky. For example, if you plan to observe the sky at 9 pm, you should select a map that shows you the stars that will appear during that hour.

 **In Real Life**

Did you know that the art and science of creating maps is known as **cartography**? It is a branch of geography.

 ## How to Use a Star Map?

Once you have been able to get a map aligned to the season and the time of your observation, you need to figure out the direction and match it to the sky. Most maps will indicate the directions north, south, east and west. First, look for star patterns. Then find the brighter stars, which will be shown as big dots. The dimmer stars will be shown as tiny dots.

A Brief History of Star Maps

The very first Western maps of the skies above the northern hemisphere and the southern hemisphere depicting stars and constellations date back to the year 1440. They are currently preserved in Vienna. It is likely that these star maps were based on two charts from 1425, which are now lost. Well-known German painter Albrecht Durer drew the first printed star maps in 1515. In 1540, Alessandro Piccolomini came out with the first book of printed star charts, called *De le stele fisse*.

Till the end of the 16th century, star charts only comprised the 48 constellations recorded by Ptolemy. Later, in 1595, Pieter Dircksz Keyser added 12 additional constellations in the southern skies. Some of them are named after birds like the toucan, peacock and the phoenix. In the 1600s, other astronomers were responsible for introducing the southern constellations through different forms of maps like globes and plates, including the one in the *Uranometria*, which is a star atlas by Johann Bayer produced in 1603.

▲ Johann Bayer's Uranometria was the first star atlas. The illustration shows the constellation of Orion on a copperplate engraving from his book

◀ These are some of the important constellations visible from the northern hemisphere in January

▶ The Farnese Globe is the oldest in the world

Isn't It Amazing!

The oldest globe in existence today is the famous Farnese Globe, which is also the oldest astronomical object of art. It dates back to the 3rd century BCE. It is perhaps a Roman imitation of a Greek globe showing constellations instead of single stars. It has been kept in the National Archaeological Museum in Naples. The very first globe was made in the 6th century BCE by Thales of Miletus.

Constellations in the Northern Hemisphere

Each of the two hemispheres have different celestial objects and constellations for you to marvel at. Some stars and constellations can be seen more clearly from the northern hemisphere. In the southern hemisphere, they would be quite low on the horizon, or not visible at all. The star map in the centre shows the constellations of the northern hemisphere.

◀ A star map showing the constellations with their perceived shapes

⭐ Polaris and the Three Circumpolar Constellations

One of the most popular and well-recognised patterns in the northern sky is that of the seven bright stars which form the Big Dipper, part of the Ursa Major or Great Bear constellation. The two stars (Dubhe and Merak) at the end of the bowl of the Big Dipper are also known as the pointer stars. If you draw a straight line through them, they will point you to Polaris or the North Star. It is located in the tail of the Little Bear or at the end of the Little Dipper in the Ursa Minor (Small Bear) constellation close by. If you extend that straight line from Polaris, you will see a constellation which looks like a 'W'. That is the constellation of Cassiopeia. Ursa Major, Ursa Minor and Cassiopeia are circumpolar constellations. The extremely bright Polaris is one of the most striking sights in the northern skies. Incidentally, Ursa Minor is the third largest constellation in the sky.

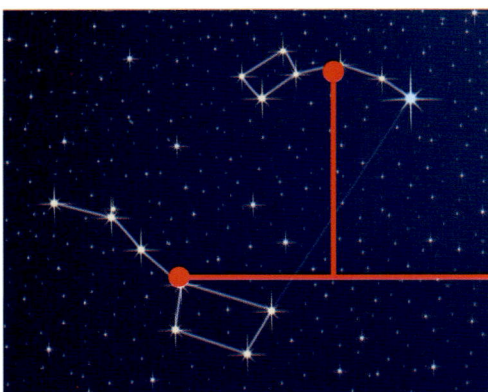

▲ Ursa Minor
▼ Ursa Major

▶ Constellations of the northern hemisphere

Myth: The Big Bear and Little Bear originate from a Greek myth. The Greeks considered the two bears to be Callisto and her son Arcas. They were turned into bears by Zeus. They could also be referring to the two bears who saved the life of baby Zeus from his cannibalistic father. The long tails of the bears were said to have been the result of Zeus swinging them far up in the sky.

Andromeda

▲ The Andromeda constellation

From Cassiopeia, if you look down to the right, you will find the Andromeda galaxy and close to it you will be able to identify a huge rectangle, which is the constellation Pegasus. Follow the left star at the base of the rectangle and you will be able to spot the Andromeda constellation. Alpheratz, which means 'the horse's navel' in Arabic, is the brightest star in this constellation. If you are lucky, you may even be able to spot a fuzzy oval in the sky, which is the amazing Andromeda Galaxy, the large galaxy closest to Earth.

Myth: Andromeda represents the princess of Ethiopia who, according to the Greeks, was saved from a sea monster by Perseus.

Isn't It Amazing!

It is believed that the luminosity of the Antares star—the brightest star in the Scorpius constellation in the Southern hemisphere—is 10,000 times greater than the Sun's luminosity.

▲ The Cepheus constellation

Cepheus

This constellation represents a king and is a dimmer constellation compared to many others. However, it is easily recognisable, being shaped like a stick house. In August and September, you will find Cepheus located on the upper right side of Polaris. The top of the roof generally points towards the Pole Star. If you are able to locate the more well-known Cassiopeia constellation, then you can spot Cepheus, which is a close neighbour. Alderamin is the brightest star in this constellation.

Myth: It is named after the King of Ethiopia and father of Andromeda. He was forced to sacrifice his daughter to a sea monster.

Constellations in the Southern Hemisphere

No study of the constellations is complete without observing those in the sky above the southern hemisphere. When viewed from the south, northern constellations will appear upside down. Also, the northern circumpolar constellations like the Big Dipper, Cassiopeia, etc., become seasonal and Polaris cannot be seen at all.

Even if you think you know your constellations well, it isn't easy to spot them after a change in hemispheres. For instance, in the southern hemisphere, the Summer Triangle becomes the Winter Triangle and similarly, some of the other constellations point in different directions, compared to how you would see them in the northern hemisphere.

⭐ Canis Major and Sirius

If you look high up into the north-eastern sky, you will be able to find Canis Major to the right (south-east) of Orion. Also known as the Greater Dog, Canis Major can proudly boast of a star called Sirius, which is the brightest star in the dark sky. It is also the fifth-nearest to Earth. The Canis Major dwarf galaxy is also part of the Canis Major constellation and is the dwarf galaxy nearest to Earth. A dwarf galaxy is small when compared to other galaxies like the Andromeda or Milky Way. It consists of about a hundred million or a couple billion stars. On the other hand, the Milky Way galaxy consists of about 200–400 billion stars.

Myth: Canis Major was considered to be Orion's hunting dog.

▲ *Canis Major with Sirius, the brightest star in the night sky*

⭐ Carina and Canopus

The second brightest star, Canopus is also found in the south in the constellation of Carina. You can spot Canopus by looking at an angle of 35° from Sirius. It can generally be viewed from October to May and is almost always seen when Sirius is visible. Earlier, Carina was a part of a larger constellation called the Argo Navis. It was divided into three different constellations—Carina, Puppis and Vela—by French astronomer Nicolas Louis de Lacaille.

Myth: Carina was earlier a part of the Argo Navis constellation, named after the ship Argo. In an old Greek myth, Jason and the Argonauts went on this ship to rescue the Golden Fleece.

◀ *Carina or the Keel constellation*

SPACE — CONSTELLATIONS

⭐ Crux or the Southern Cross

The Southern Cross is the most familiar and well-known pattern in the southern hemisphere. It is the most striking feature of the Crux constellation with its five bright stars roughly forming the shape of a cross. Crux is the smallest constellation amongst all the constellations. Two of its brighter stars are Acrux and Gacrux, which point towards the southern celestial pole. A dark nebula, the Coalsack Nebula is also part of this constellation.

◀ *Crux or the Southern Cross*

⭐ Alpha Centauri: The Nearest Star System

It will be well worth your while to visit the southern hemisphere in order to see the closest star system to Earth, the Alpha Centauri. It also happens to be the third brightest star system in the night sky. Being circumpolar, you can see Alpha Centauri all through the year if you live south of the Equator. Sometimes in May, Alpha Centauri can be seen a few degrees above the southern horizon. It is part of the constellation Centaurus.

▲ *The Alpha Centauri star system is about 4.3 light years from Earth*

▲ *Constellations in the southern hemisphere*

👤 In Real Life

Interestingly, the five stars of the Crux appear on the flags of countries like New Zealand, Papua New Guinea, Samoa, Australia and Brazil. They are also part of the national anthems of the latter two countries.

▶ *The Australian and Papua New Guinean flags depict stars from the Southern Cross*

Finding Stars & Constellations

Wouldn't it be fun if you could find and recognise stars and constellations in the night sky? One way of spotting them is by getting familiar with their names and remembering their shapes and relative sizes. You can begin by trying to identify constellations which can be seen by the naked eye. Some of them are Orion, Aquarius, Aquila, Aries, Canis Major, Cygnus (or Northern Cross), Leo and Scorpius.

⭐ I Spy Cassiopeia, the Queen

An easy to find constellation consisting of five bright stars, roughly in the shape of a 'W', is Cassiopeia the Queen. Another way of spotting it is to see five stars that make up one narrow angle and one wide angle. Like Ursa Major, it rotates around the north celestial pole, and so it may appear as an upside-down 'M', during some parts of the year!

Myth: Cassiopeia was a beautiful but vain Queen of Ethiopia. To punish her, Poseidon ordered that her daughter Andromeda be killed at the hands of a sea monster. Perseus managed to save Andromeda, but Cassiopeia plotted to kill him. To avenge himself, Perseus turned Cassiopeia into stone.

▲ *Cassiopeia or the Queen constellation in the northern sky during winter*

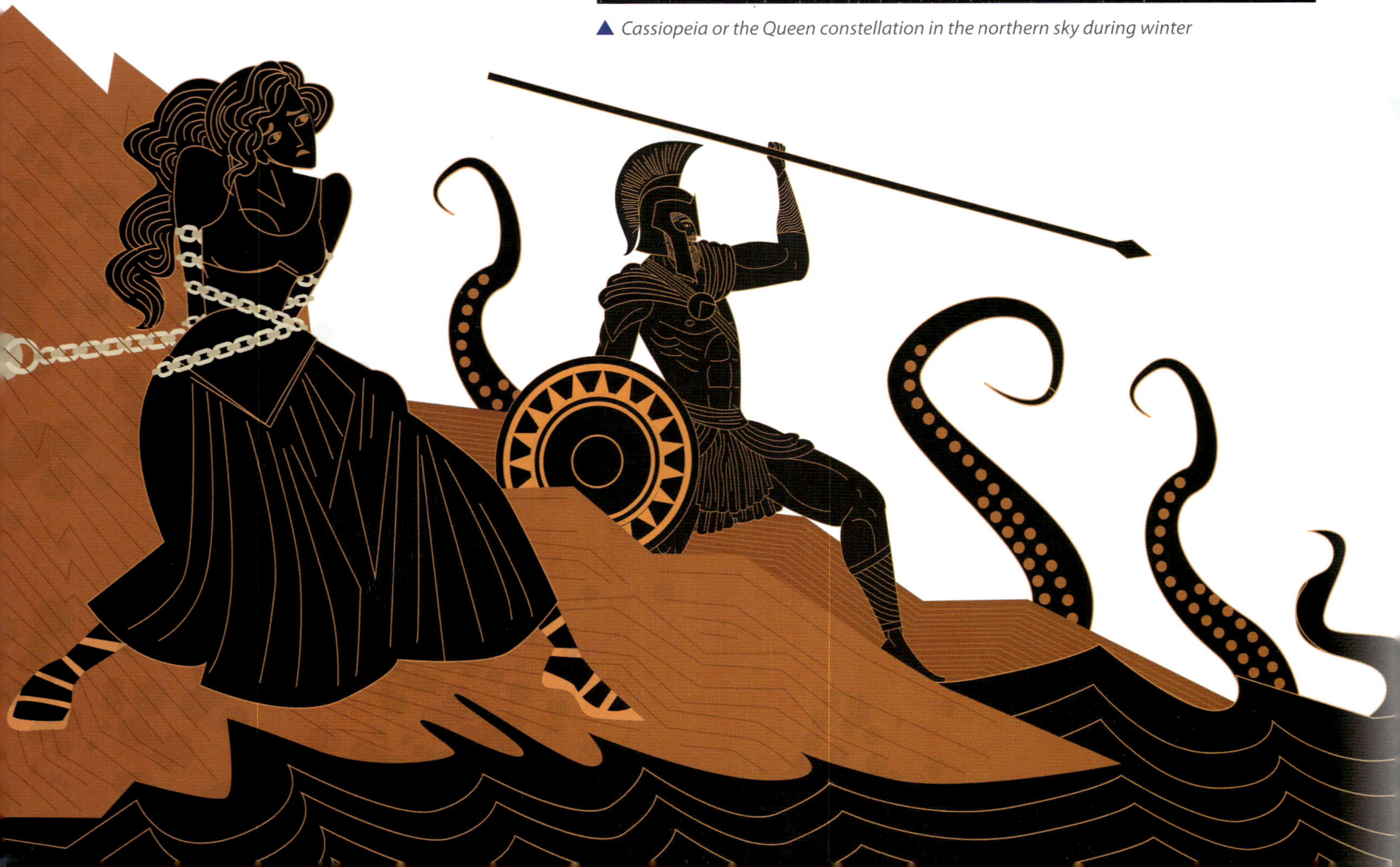

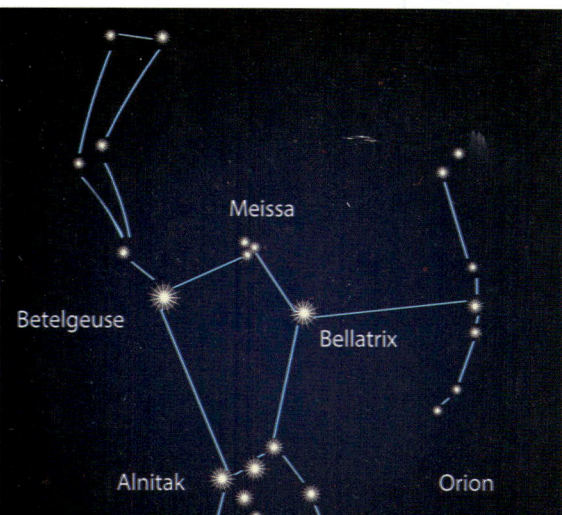

 I Spy Orion, the Hunter

Orion is one of the easiest constellations to spot if you can see its distinctive belt formed by three stars in a straight line. The constellation is roughly in the shape of the letter 'H'. The shoulders have one star each. The first is Betelgeuse (meaning 'armpit of the giant') and the other is Bellatrix (meaning 'warrior'). He also has two stars in the feet, called Rigel and Saiph. From the belt, if you draw a line to the left, you will come to the star Sirius. Orion is usually seen in the standing position.

Myth: Orion was a brave and mighty hunter who fell in love with Artemis, the Moon Goddess. Out of jealousy, her brother Apollo sent Scorpius (the scorpion) to fight with Orion. Both died during the fight. It is said that Zeus put Orion in the winter sky and Scorpius in the summer sky so that they could not fight any more.

◀ *Orion, the mighty hunter, with a few of its main stars*

 I Spy Taurus, the Bull

To spot this constellation, look for the horns of the Bull which form a 'V' shape and try to find his bright red eye, which is the Aldebaran star.

Myth: In ancient Greece, Zeus was considered to be the ruler of heaven. According to legend, he kidnapped Europa on the sly by turning himself into a large white bull and hid her on the island of Crete, where she bore him numerous children.

By learning techniques to spot Orion, Cassiopeia and Taurus, each differently shaped (H, W and V) and sized, you might have got the hang of recognising constellations in the sky!

▼ *Taurus, the Bull constellation*

Isn't It Amazing!

The Serpens or Serpent constellation is the only constellation divided into two parts—Serpens Caput, which refers to the head of the serpent, and Serpens Cauda, meaning tail of the serpent. They are both part of the serpent that Ophiuchus the giant seems to be holding.

▲ *An illustration of the Serpens constellation*

The 'C' Constellations

Interestingly, there are 22 constellations which begin with the letter 'C'. Some of them are the Corona Borealis, the Crater, the Corvus and so on.

⭐ Corona Borealis

Corona Borealis, also called the Northern Crown, is found in the northern hemisphere. It is fairly small, in fact it ranks 73rd in size in the total list of 88 constellations. Corona Borealis is in the shape of an open and irregular semicircle. It has four bright stars and lies in between the constellations of Boötes and Hercules. Alphecca is the brightest star in this constellation.

Myth: It gets its name from the crown that the Greek God Dionysus gave to Ariadne, the Cretan princess.

▲ *In Latin, Corona Borealis means the Northern Crown*

⭐ Crater

Crater is a constellation in the southern sky, also called the Cup. It is quite faint and ranks 53rd in terms of its size. Crater has three stars with planets and its brightest star is Delta Crateris.

Myth: This constellation is connected to Corvus (Crow) and Hydra (Water Snake). The Greek God Apollo sent a crow to bring a cup of water for a ritual. The crow got distracted and failed to bring the cup of water to him. Instead, it returned with a water snake and blamed it. Apollo got angry and threw all three—the crow, the cup and the water snake—up into the sky.

▲ *The Crater or Cup constellation*

⭐ Corvus

Corvus means 'crow' or 'raven' in Latin. This is a constellation from the southern hemisphere. Hydra, Virgo and Crater are its neighbouring constellations. In size, it is the 70th constellation. The brightest star in Corvus is Gienah.

Myth: In Greek mythology, Corvus is the sacred bird of Apollo.

▲ *Corvus or the Crow constellation*

💡 Isn't It Amazing!

The Corvus constellation is 60 million **light years** away. Corvus houses two large galaxies which have collided, causing many new stars to be formed. Since these colliding galaxies look like the arcs of an antenna, they have been named the Antennae galaxies.

▲ *The two large Antennae galaxies (NGC 4038 and NGC 4039) in Corvus. The picture on the right shows when they collided*

Legends of the Horses

Two well-known constellations are associated with fascinating creatures such as a winged horse and a centaur. The Pegasus constellation is in the shape of a winged horse, while the Centaurus constellation is in the shape of a centaur—a creature that is half-horse and half-man.

Pegasus or the Winged Horse

This constellation is one of the biggest in the northern sky and also the seventh-largest overall. It is best known for the 'Great Square' of Pegasus, a well-recognised asterism in that part of the hemisphere. Pegasus is also known for some of its bright stars and **deep-sky objects**.

Some of the constellations nearby are Andromeda, Aquarius (Water Bearer), Cygnus (Swan), Delphinus (Dolphin), Equuleus (Little Horse), Lacerta (Lizard), Pisces (Fishes) and Vulpecula (Little Fox).

Myth: Pegasus—or the winged horse—was born from the blood of Medusa's severed head. Bellerophon was a Greek hero who captured Pegasus and rode on him during his fight with a fire-breathing female monster. He tried to flee to heaven on Pegasus's back but was killed (some accounts say he was wounded) during the escape. The horse became a constellation and a servant of Zeus.

▲ *The Pegasus constellation*

▲ *Pegasus*

Centaurus

Centaurus lies in the southern sky and is the ninth-largest constellation. It is home to the two stars nearest to Earth—Proxima and Alpha Centauri (also the fourth-brightest in the sky). It also contains Beta Centauri, the 11th-brightest star.

Myth: This constellation is linked with the centaur Chiron. Chiron was wise and peaceful, unlike most other centaurs who were aggressive. According to lore, it was Chiron who taught other Greek heroes and invented the constellations.

▼ *A centaur's rough outline traced in the Centaurus constellation*

💡 Isn't It Amazing!

The constellation Draco or Dracon is named after a Greek lawgiver from the 7th century BCE. Draco was known to give very severe punishments in Athens, and often gave the death sentence. The word 'draconian', which originated from his name, is now commonly used to describe extremely strict and oppressive laws.

A Tale of Two Giants

There are two giants among the constellations. One is named after a figure known for his strength and valour; the other is a constellation that is notorious for being left out from the zodiac constellations, even though it lies in the same region as the others.

Hercules—the Giant

Hercules, sometimes referred to as the Kneeling Giant, is the fifth-largest constellation. It is situated between two bright stars—Arcturus (in the Boötes constellation) and Vega (in the Lyra constellation). A square figure in the middle of the Hercules constellation, known as the Keystone, is one of its more noticeable aspects. Besides Boötes and Lyra, some of the other neighbouring constellations are Aquila (Eagle), Corona Borealis, Draco, Lyra (Lyre), Ophiuchus, Sagitta (Arrow), Serpens Caput (Head of the Serpent) and Vulpecula (Little Fox).

▲ *Plate 11 of Urania's Mirror, a set of cards illustrating various constellations, depicts Hercules and Corona Borealis*

Myth: In the 5th century BCE, Hercules was first identified as Heracles, a well-known Greco-Roman hero and the son of Zeus borne by Alcmene. Zeus was eager for Hercules to become the ruler of Greece, but his jealous wife Hera tricked him and got her sickly son to become king. Hercules grew up and suffered at the hands of Hera. His first victory was as an infant, when he slayed two serpents that Hera had sent to kill him. He is often depicted holding a club and wearing lion skin.

◄ *A star chart showing Hercules, the constellation referred to as the Kneeling Giant*

Ophiuchus—the Other Giant

The name 'Ophiuchus' has Latin origins and refers to a serpent-bearing man. This constellation stands out for several reasons. The 'feet' of this constellation overlap with a part of the Scorpius constellation. It is located on the **ecliptic**. When the Moon crosses the ecliptic, we see solar or lunar eclipses. Ophiuchus has not been included in the 12 zodiac constellations lying in the same region. The second-nearest star to the Sun, the Barnard's Star is located in this constellation.

Myth: Ophiuchus (son of Apollo, the God of healing, truth and prophecy) symbolises the Greco-Roman God of medicine and is seen holding a serpent, which is considered to be a symbol of renewal.

▲ *In some cultures, the Ophiuchus constellation is considered to be the 13th zodiac constellation*

Water Creatures

There are 10 water creatures represented in the long list of constellations. Two of the important water constellations are shaped like a water snake and a clever dolphin.

⭐ Hydra—the Water Snake

Hydra is a constellation seen in the southern hemisphere. This constellation is the largest in the night skies. It consists of several ordinary stars, the brightest of which is Alphard, whose name means 'brightest star' in Greek, as it is derived from the word 'alpha'. The constellation seems to have been created to mark the Equator in around 2800 BCE. In Greek mythology, the number of Hydra's heads changes from 5–100 depending on who is telling the story. However, the Hydra constellation has only one head.

Myth: Hydra is linked to the Corvus (Crow) and Crater (Cup) constellations. The water snake was thrown into the sky by Apollo, along with the cup and the crow due to its disobedience.

▲ The constellation of Hydra

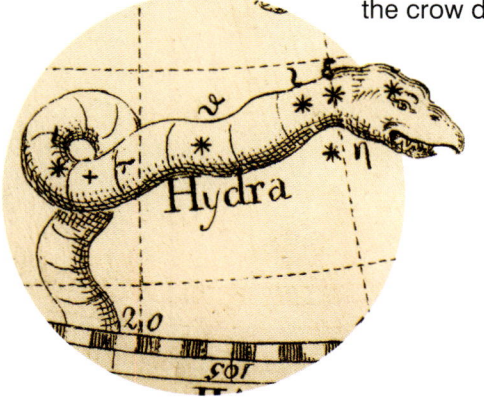
▲ Hydra—the water snake

Incredible Individuals

Although the Southern Cross or Crux has been written about since ancient times, the first to document it as a constellation, in a set of star maps in 1679, was the French architect and cartographer, Augustin Royer. He also added two more constellations. Royer split some part of the Canis Major constellation and named it the Columba (Dove) by including some stars from the Centaurus constellation in it.

⭐ Delphinus—the Dolphin

Delphinus is a small constellation in the northern hemisphere. Rotanev is its brightest star. It consists of an asterism called 'Job's Coffin', which is shaped like a diamond.

Myth: Delphinus has two mythological roots. He was the messenger of Poseidon, the Sea God. In another account, he was the dolphin that saved the poet Arion from drowning.

▲ Delphinus or the Dolphin constellation

In Real Life

We know about air, water and land pollution, but have you heard about **light pollution**? It hampers our view of the night skies. Hong Kong is believed to be one of the most light-polluted cities in the world. In the last 100 years or so, light pollution has made it very difficult for stargazers and astronomers alike to study the skies due to the intense glare it creates. Organisations like the International Dark-Sky Association have been fighting against light pollution and struggling to preserve the dark skies for future generations. They support and propose 'smart lighting' designs.

The Zodiac

There are 12 constellations which lie in a region of the sky known as the zodiac. The word 'zodiac' comes from ancient Greek. Since most of the constellations in this area represented animals, the Greeks called it *zodiakos kyklos* (circle of animals), or 'ta zodia' (the little animals).

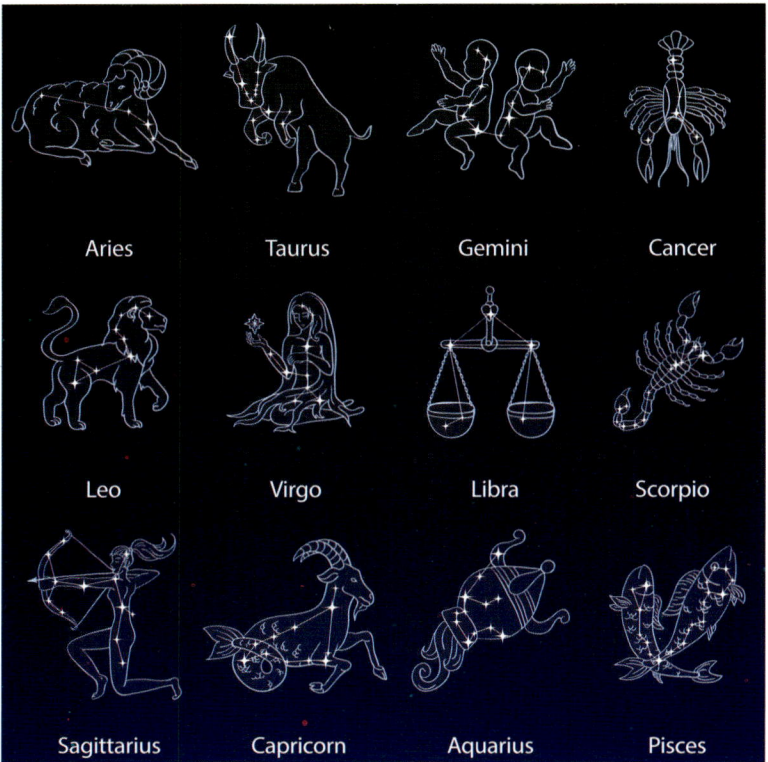

◀ The 12 zodiac constellations with their shapes and names

▲ The zodiac constellations are mostly shaped like animals but some feature human figures and objects

⭐ What is the Zodiac?

In astronomy and astrology, a particular belt or area in the sky extending 9° on both sides of the ecliptic is known as the zodiac region. The orbits of the Moon and the main planets also lie completely inside this region.

It was the early astronomers who noticed that the Sun travels through the signs of the zodiac throughout the year and stays for about a month in each of them. Think of a straight line that is drawn from Earth, passing through the Sun and moving out beyond our solar system to where the bright stars are. As Earth orbits around the Sun, the line also rotates and points to different stars over a period of a year, which is one complete rotation. The constellations through which the line passes are in the zodiac.

These 12 constellations lying in the zodiac belt are known as the zodiac constellations. For believers of astrology, each of these correspond to and govern a period in the calendar. So, for example, a person born between March 21 and April 19 will have Aries as their zodiac sign. Each of the other 11 signs are similarly associated with specific periods and dates.

⭐ Why is Aries Considered to Be the First Zodiac Sign?

Mid-March (around the 20th or 21st) marks the beginning of spring. It brings relief after a harsh winter and is an important time in agriculture. Springtime is hence celebrated in several cultures all over the world. Almost 4000 years ago, even ancient Babylonians celebrated this period to mark this rebirth and the beginning of a new year. Similarly, in 600 BCE, people living in Iran celebrated Nowruz (New Day), most likely as part of the Zoroastrian religion. The Romans considered the beginning of spring as the start of the year. It is for this reason that Aries, the zodiac representing the beginning of spring, is considered to be the first constellation or sign of the zodiac. It reflects the importance that various cultures ascribed to routine astronomical phenomena.

Aries: the Ram

Location: Northern sky between Pisces and Taurus

Period: March 21–April 19

Brightest star: Although the Aries constellation does not have any bright stars, the brightest star within it is called Hamal, which means 'sheep' in 'Arabic'.

▲ The Aries constellation

Taurus: the Bull

Location: Northern sky between Aries and Gemini

Period: April 20–May 20

Brightest star: Aldebaran (meaning 'the follower' in Arabic) is the brightest star in the Taurus constellation. It is the 14th-brightest star found in the sky.

▲ The Crab nebula, the Pleiades and the Hyades star clusters are also found within the constellation of Taurus

Gemini: the Twins

Location: Northern sky between Cancer and Taurus

Period: May 21–June 21

Brightest stars: Castor and Pollux are the brightest stars in the Gemini constellation. Pollux is brighter than Castor and is the 17th-brightest star in the sky.

▲ The Gemini constellation contains Geminga, an isolated pulsar

Cancer: the Crab

Location: Northern sky between Leo and Gemini

Period: June 22–July 22

Brightest star: Al Tarf (meaning 'the end' in Arabic. It indicates the end of one of the crab's legs), although a fairly dim star, it is the brightest star in the Cancer constellation.

▲ The well-known star cluster known as Praesepe or the Beehive is found in the constellation of Cancer or the Crab

Leo: the Lion

Location: Northern sky between Cancer and Virgo

Period: July 23–August 22

Brightest star: Regulus, which means 'little king' in Latin. It is also known as Alpha Leonis, and is one of the brightest stars in the night skies.

▲ A lot of the stars comprising the Leo constellation form an asterism known as the Sickle

Virgo: the Virgin

Location: Southern sky between Leo and Libra

Period: August 23–September 22

Brightest star: Spica, which means 'head of grain' in Latin. It is also referred to as Alpha Virginis. It is the 15th-brightest star in the entire sky.

▲ The Virgo cluster, the closest large cluster of galaxies, is located in the constellation of Virgo

Libra: the Balance

Location: Southern sky between Scorpius and Virgo

Period: September 22–October 23

Brightest star: The stars in Libra are faint, but amongst them, the brightest is Zubeneschamali. The name means 'northern claw' in Arabic.

▲ Libra is represented by a balance scale or a woman holding a balance scale

Scorpius: the Scorpion

Location: Southern sky between Libra and Sagittarius

Period: October 24–November 21

Brightest star: Antares, also called Alpha Scorpii, is the brightest star in the Scorpius constellation. It is a large star much bigger than the Sun.

▲ Within the constellation of Scorpius lies Scorpius X-1, the brightest source of X-rays in the sky

Sagittarius: the Archer

Location: Southern sky between Capricornus and Scorpius

Period: November 22–December 21

Brightest star: Kaus Australis is the brightest star in the Sagittarius constellation. The word 'Kaus' means 'bow' in Arabic and 'Australis' is 'southern' in Latin.

▲ Many stars in the constellation of Sagittarius form the well-known asterism known as the Teapot

Capricornus: the Goat

Location: Southern sky between Aquarius and Sagittarius

Period: December 22–January 19

Brightest star: This constellation's stars are faint but Deneb Algedi (meaning 'kid's tail' in Arabic) is the brightest amongst them.

▲ Capricornus means 'goat-horned' in Latin

Aquarius: the Water Bearer

Location: Southern sky between Capricornus and Pisces

Period: January 20–February 18

Brightest star: Sadalmelik or Alpha Aquarii is the brightest star in the Aquarius constellation. It means 'lucky stars of the king' in Arabic.

▲ Aquarius, the Water Bearer

Pisces: the Fishes

Location: Northern sky between Aries and Aquarius

Period: February 19–March 20

Brightest star: Eta Piscium refers to the cord that connects the two fishes in the Pisces constellation.

▲ The Pisces constellation has only dim stars with no outstanding grouping

What is Astrology?

Astrology is the practice of interpreting the movements of celestial bodies to predict their influence on human life. Since ancient days, human beings have observed and interpreted the Sun, the Moon, the planets and stars in order to try and figure out their impact on people, events and life on Earth.

▲ *An astrological clock with zodiac symbols, in Prague*

Astrology is not the Same as Astronomy

Astrology and astronomy are completely different. Astronomy is the scientific study of everything in outer space. Astronomers and other scientists know, for a fact, that the stars, the Sun, the Moon and planets have no influence on the day-to-day activities and lives of human beings. Astrology, on the other hand, is unscientific; nobody can prove its impact on people. Although astrology is based on superstition, it is still a popular part of multiple cultures. This shows that many civilisations were able to spot these astonishing patterns of stars in the skies and attributed various meanings to them. Humans often looked to the heavens for phenomena that science could not explain at the time.

⭐ The Zodiac and Astrology

A **horoscope** is a detailed chart that supposedly shows what impact celestial bodies have on the life of a person born at a particular time. The zodiac sign under which a person is born is an important consideration in such predictions, as is the position of the planets. Each of the zodiac constellations is considered to be the 'home' of one or more planets. The zodiacs are believed to have a strong or weak influence on the person. However, such an influence or correlation does not exist within the realms of modern science. It is very important to learn to differentiate between science and superstition, and not fall prey to any misleading claims.

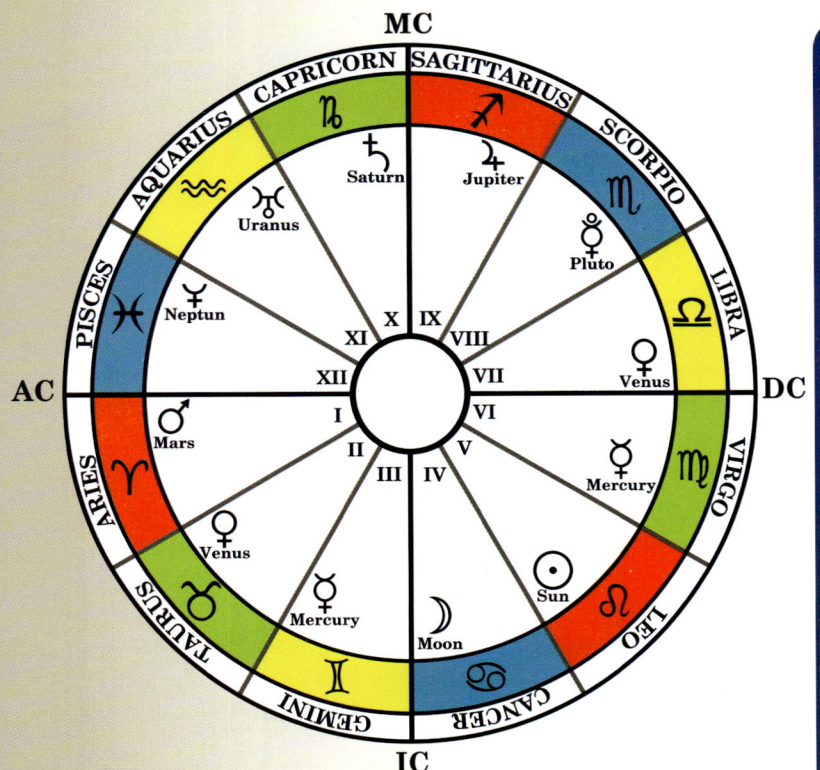

▲ *A birth chart with zodiac signs and the planets that mythically govern each of them*

🏅 Incredible Individuals

American astronomer Benjamin Apthorp Gould (1824–1896) was a child prodigy with exceptional abilities. When he was just three years old, Gould was able to read aloud, and at age five, he could compose Latin poems! One of Gould's most significant contributions as an astronomer was his work on star catalogues, which helped fix the list of constellations in the southern hemisphere. Gould had studied mathematics and the physical sciences at Harvard University. In 1848, he became the first American to earn a doctorate degree in astronomy from the University of Göttingen in Germany. In 1849, he started *The Astronomical Journal*, the first of its kind in the US, based on professional astronomical research.

▶ *Benjamin Apthorp Gould*

Chinese Constellations

Astronomy had a royal role to play in ancient Chinese culture. Emperors hired astronomers to record astronomical events, precisely record time and calculate the calendar. Chinese astronomers developed their own methods and ways, which were different from the ones used in Europe and elsewhere. Rather than theorising, they were more interested in making improvements in the accuracy of their measurements and in individual events like sightings of comets, novae, meteor showers, solar eclipses and sunspots (the last of these were first discovered by the Chinese, long before the Europeans).

Imperial Importance of Chinese Constellations

Stars and constellations were particularly significant to the Chinese, since they believed in the harmony of the heavens, Earth and human beings. In fact, events within Chinese dynasties were closely correlated to the stars. The Emperor was considered to be the 'Son of Heaven' and was supposed to have been given heavenly instructions to rule the country. Constellations were also important in Chinese culture because most of them were linked to the hierarchy within the dynasties on Earth and were considered to be a replication of the same in the heavens. By around 220 CE, which marked the end of the Han Dynasty, Chinese court astronomers had already named 283 constellations by grouping 1,464 stars. Predicting events accurately and with attention to detail was important so that they could advise the Emperor.

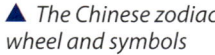
▲ *The Chinese zodiac wheel and symbols*

3 Enclosures, 4 Symbols and 28 Mansions

The Chinese had a very different method of naming stars and constellations. They considered the north celestial pole to be the centre of heaven since the stars revolved around it. This area around the pole was segregated into three enclosures or sectors—the Purple Forbidden Enclosure (紫微垣, Zǐ Wēi Yuán); the Supreme Palace Enclosure (太微垣, Tài Wēi Yuán) and the Heavenly Market Enclosure (天市垣, Tiān Shì Yuán). These sectors represented the organisation of the dynastic hierarchy on Earth. The Purple Forbidden Enclosure was the most important of them all.

Chinese astronomers also assigned four symbols spread in the region of the ecliptic zodiac and the lunar orbit, which corresponded to the four cardinal directions. Each symbol was represented by an emblem—The Azure Dragon for the east; the White Tiger for the west; the Vermilion Bird for the south; and the Black Tortoise for the north. Each of these four symbols were divided into seven sections, which were known as mansions. There are 28 mansions in total. Most of the important Chinese constellations are situated within these mansions.

Incredible Individuals

Luoxia Hong was a Chinese astronomer during the Han dynasty. Hong was well-noted for introducing a new calendar in the court of Emperor Wu-ti, who had invited proposals from astronomers across his empire.

In ancient China, it was believed that the emperor or ruler was given the right to rule by the heavens. When there was a change of emperor, and particularly if there was a change in the ruling dynasty, changing the calendar was not just an obligation of the ruler, but it was done to show the emperor's link with the heavens. The new calendar served to establish a new regime with new heavenly influences.

The calendar developed by Hong and his colleague Deng Ping was selected as the best from among 18 others by emperor Wu-ti. Hong's calendar was put to use in 104 BCE and incorporated both the Sun and the Moon into a common system. Hong also predicted eclipses and the positions of the planets based on accurate observations which were possible due to a special equipment he used for this purpose.

SPACE | CONSTELLATIONS

Dragon: East	Tiger: West	Vermilion Bird: South	Tortoise: North

角 (Jio) Horn	奎 (Kuí) Legs	井 (Jng) Well	斗 (Du) Dipper
亢 (Kàng) Neck	娄 (Lóu) Bond	鬼 (Gu) Demon	女 (N) Woman
氐 (D) Root	胃 (Wèi) Stomach	柳 (L) Willow	牛 (Níu) Ox
房 (Fáng) Room	昴 (Mo) Hairy Head	星 (Xng) Star	虚 (X) Emptiness
心 (Xn) Heart	毕 (Bì) Net	张 (Zhng) Growth	危 (Wi) Danger
尾 (Wi) Tail	觜 (Zu) Turtle Beak	翼 (Yì) Wings	室 (Shì) Room
箕 (J) Winnowing basket	参 (Cn) Three Stars	軫 (Zhn) Deep emotion	壁 (Bì) Wall

▼ *Made in 1439, an armillary sphere at the Beijing Ancient Observatory*

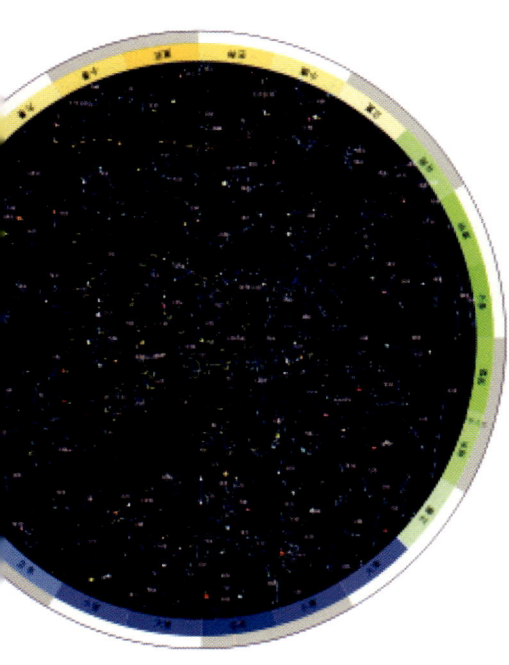

▲ *A modern star chart with the 28 mansions indicated on the border of each hemisphere*

Stargazing Apps & Astrotourism

With the advent of the Information and Digital Age, you no longer need to lug around a telescope or an unwieldy star chart to admire your favourite stars and constellations. Your mobile phone can serve as your window to the world of stars! Several mobile apps are available as an aid to stargazing. They help spot the twinkling trinkets in the sky and also send you an alert about upcoming stargazing events. Astrotourism is also picking up. People are seeking destinations with dark, unpolluted skies and clearly visible stars.

▶ Mobile phone applications are well-suited for amateur astronomers and youngsters

 ## Identifying Celestial Objects

There are various apps for iOS and Android that are easy and simple to use. All you need to do is hold your phone up and point it at the night sky. The app will show you which stars, planets and constellations you are looking at. If something of interest catches your eye, you can tap the screen and get additional information about those celestial objects.

 ## Map Your Stars

Other apps show detailed maps of stars, galaxies and planets on the screen along with information about what you are seeing when you hold up your phone or tablet. There are settings for enlarging the view and animated graphics. You can even check out what the sky will be like on a future date. There are also added companion watch apps that reveal phases of the moon and a chance to spot space stations, etc.

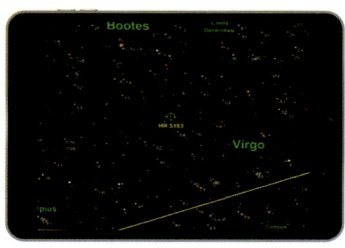

▲ The latest apps even offer 3D animated graphics of planets and other celestial objects

⭐ Different City Views

Another interesting feature offered by some apps is the traveller mode which allows you to see what the sky looks like from different cities.

▲ Advancements in technology have made it possible for us to discover the marvels of space with an app

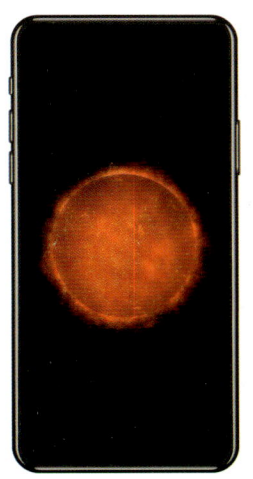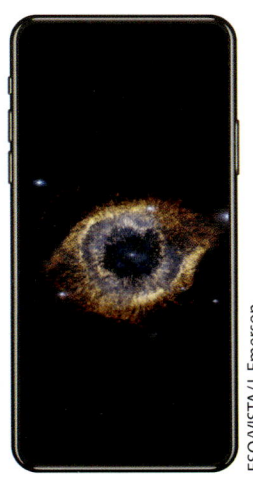

◀ Apps even offer fun means of learning more about space, such as informative and interesting movies

⭐ Movies and More

Some apps allow you to view models of satellites, spacecrafts and space missions, track their paths and even view photos taken on these missions. There are apps with detailed 2D and 3D graphics, animations and movies about a wide range of subjects like how a red dwarf star works, the history of our Moon, black holes, etc. They are easy to understand even for a person without much of a background in science.

⭐ Astrotourism

Astrotourism has picked up in several countries including USA, Canada, Australia and others. It involves visits to places of astronomical interest and importance, like planetariums and observatories. Astrotourists seek out places where stars can be easily seen in unpolluted, dark skies. It is not just limited to stargazing and also involves taking trips to watch an eclipse, a meteor shower, the Northern Lights, a rocket launch or space-related experience. The International Dark-Sky Association in the US has listed over 60 such International Dark Sky Parks which people can visit. Most of these parks are generally near ski resorts, mountains or state and national parks. Stargazing events, parties and safaris are also gaining popularity now.

Our phones often distract us and keep us away from many wonderful things—stars and constellations for example. Like poet W. H. Davies said, "What is this life if, full of care, we have no time to stand and stare."

▲ The Belgrade Planetarium in Serbia is an ideal place to visit for astrotourism.

👤 In Real Life

In June 2018, National Geographic along with Au Diable Vert Mountain Station, launched ObservEtoiles which is the world's first open-air **augmented reality** planetarium in Glen Sutton, Quebec, Canada.

Official Naming of Constellations

In 1919, the International Astronomical Union (IAU) was set up to promote and safeguard astronomy. This included research, education and development, communications, etc. IAU is recognised worldwide, and it is the only 'official' organisation allowed to name celestial bodies in the sky. More than 107 countries are members of the IAU.

IAU and the Constellations

Constellations were earlier informally described using the approximate shapes formed by the stars located in them. With an increase in the number of space-related discoveries in the 20th century, there arose a need to define them by specific boundaries. This was important while naming variable stars, for instance. Variable stars keep changing, becoming brighter or fading, instead of shining steadily like regular stars. Such stars are easier to spot by the constellations in which they are located. Hence, it was important to have a universally approved boundary for each constellation.

Setting Boundaries

In 1922, the IAU officially adopted the list of 88 constellations that we use today. Definitive boundaries between constellations, which extend out beyond the star figures, were set in 1930, so that every star, nebula, and galaxy, now lies within the limits of one constellation. For today's astronomers, constellations refer not so much to the patterns of stars, but to precisely defined areas of the sky. The IAU established three-letter abbreviations for each constellation. For example, Andromeda's abbreviation is 'And' and Draco's is 'Dra'.

In Real Life

One of the best places to stargaze is Hanle, in Ladakh, India. The Indian Astronomical Observatory (IAO) here has one of the highest optical telescopes in the world. It is a remote region with very little rainfall, hardly any snow and minimal cloud cover. So, these conditions make it the ideal location for astronomical observations. Moreover, being 4 kilometres above sea level is an added advantage as scientists are able to collect a wider range of data compared to a location at a lower level. The IAO also has an area for gamma-ray telescopes.

▼ The Indian Astronomical Observatory in Hanle, Ladakh

NASA's New Constellations

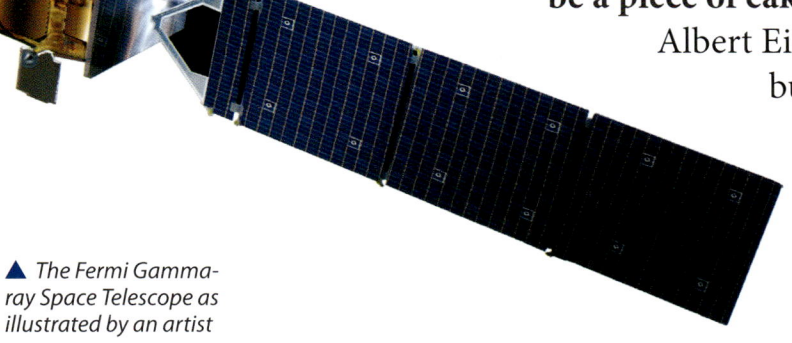

▲ *The Fermi Gamma-ray Space Telescope as illustrated by an artist*

Spotting the Great Bear, or the Orion constellation may now be a piece of cake for you. But can you spot the Hulk or Albert Einstein constellations? You may try hard, but you will need more than just your eyes to see them. You will need an extremely powerful and special telescope to view NASA's 21 new unofficial constellations.

High-energy Sky Constellations

NASA scientists came up with some new constellations in October 2018 to celebrate the completion of a decade of its Fermi Gamma-ray Space Telescope Mission. In fact, it is Fermi which was responsible for spotting these constellations which have been developed from sources of **gamma rays** in the sky. Gamma rays are a type of **electromagnetic radiation** which have the shortest **wavelength**.

Human beings can only see visible light when they look up into the sky, other sources of light like gamma rays are invisible. Since 2008, NASA's Fermi Gamma-ray Space Telescope has been scanning the heavens and observing different sources of gamma rays, the highest-energy light in the universe. Gamma-ray emissions come from various cosmic phenomena. Since Fermi began operations, it has managed to map and measure more than 3,000 different sources of gamma rays.

Hall of Fame

These gamma-ray constellations have been named after modern characters, scientists and famous landmarks. Others are related to some scientific ideas or tools.

Some of them include famous characters and objects such as the Little Prince, Godzilla, the spaceship USS Enterprise from Star Trek, Sweden's recovered warship, Vasa, the Washington Monument, Mount Fuji of Japan, Schrodinger's Cat, Radio Telescope, the Eiffel Tower, the Golden Gate and the Black Widow Spider.

▲ *View NASA's fun but unofficial constellations. Screen grab from https://fermi.gsfc.nasa.gov/science/constellations*

A Collaborative Effort

NASA was able to develop the Fermi Gamma-ray Space Telescope in collaboration with the US Department of Energy. NASA has also partnered with several academic institutions and other partners in Italy, Japan, France, Germany, Sweden and America for this mission.

Word Check

Astrology: It is the practice of interpreting the movements and relative positions of celestial bodies and their alleged influence on human affairs and the natural world. It does not have any scientific basis.

Astronomical map: It is a scientifically drawn chart of stars, galaxies, surfaces of planets and the Moon in the night sky.

Augmented reality: In computer programming, it is a process of combining or 'augmenting' (making greater in size or value) video or photographic displays by overlaying the images with useful computer-generated data.

Cartography: It is the science of drawing maps.

Circumpolar constellations: They are the constellations that never set below the horizon when seen from a particular location on Earth.

Constellations: They refer to the groups of stars that form an imagined pattern in the sky. These constellations are imagined by those who named them to form conspicuous configurations of objects or creatures in the sky.

Deep-sky objects: They are celestial objects that exist outside our solar system.

Ecliptic: It is the great circle representing the apparent path of the Sun among the constellations in the course of a year.

Electromagnetic radiation: In physics, it is defined as the flow of energy at the universal speed of light through space or through a material in the form of electric and magnetic fields that make up electromagnetic waves such as radio waves, visible light and gamma rays.

Gamma rays: They are a type of electromagnetic radiation which have the shortest wavelength.

Horoscope: It is a birth chart created according to theories of astrology.

Light pollution: It is the inappropriate or excessive use of artificial lights on Earth.

Light year: It is the distance travelled by light moving in a vacuum in the course of one year.

Mythological: It refers to something that is based on a myth or legend.

Pulsars: They are pulsating radio stars. These cosmic objects were first discovered because they had an extremely regular pulse of radio waves. Other such similar objects sometimes also emit short rhythmic bursts of visible light, X-rays and gamma radiation; and some others are 'radio-quiet'.

Wavelength: It is the distance between corresponding points of two consecutive waves.